MIX
Papier aus verantwortungsvollen Quellen
Paper from responsible sources
FSC® C105338

Dr. B. Vinoth Kumar
Dr. K.N. Vijeyakumar
K. Saranya

Single Precision Floating Point Multiplier

Anchor Academic Publishing

Vinoth Kumar, B., Vijeyakumar, K.N., Saranya, K.: Single Precision Floating Point Multiplier, Hamburg, Anchor Academic Publishing 2017

Buch-ISBN: 978-3-96067-155-8
PDF-eBook-ISBN: 978-3-96067-655-3
Druck/Herstellung: Anchor Academic Publishing, Hamburg, 2017

Bibliografische Information der Deutschen Nationalbibliothek:
Die Deutsche Nationalbibliothek verzeichnet diese Publikation in der Deutschen Nationalbibliografie; detaillierte bibliografische Daten sind im Internet über http://dnb.d-nb.de abrufbar.

Bibliographical Information of the German National Library:
The German National Library lists this publication in the German National Bibliography. Detailed bibliographic data can be found at: http://dnb.d-nb.de

All rights reserved. This publication may not be reproduced, stored in a retrieval system or transmitted, in any form or by any means, electronic, mechanical, photocopying, recording or otherwise, without the prior permission of the publishers.

Das Werk einschließlich aller seiner Teile ist urheberrechtlich geschützt. Jede Verwertung außerhalb der Grenzen des Urheberrechtsgesetzes ist ohne Zustimmung des Verlages unzulässig und strafbar. Dies gilt insbesondere für Vervielfältigungen, Übersetzungen, Mikroverfilmungen und die Einspeicherung und Bearbeitung in elektronischen Systemen.

Die Wiedergabe von Gebrauchsnamen, Handelsnamen, Warenbezeichnungen usw. in diesem Werk berechtigt auch ohne besondere Kennzeichnung nicht zu der Annahme, dass solche Namen im Sinne der Warenzeichen- und Markenschutz-Gesetzgebung als frei zu betrachten wären und daher von jedermann benutzt werden dürften.

Die Informationen in diesem Werk wurden mit Sorgfalt erarbeitet. Dennoch können Fehler nicht vollständig ausgeschlossen werden und die Diplomica Verlag GmbH, die Autoren oder Übersetzer übernehmen keine juristische Verantwortung oder irgendeine Haftung für evtl. verbliebene fehlerhafte Angaben und deren Folgen.

Alle Rechte vorbehalten

© Anchor Academic Publishing, Imprint der Diplomica Verlag GmbH
Hermannstal 119k, 22119 Hamburg
http://www.diplomica-verlag.de, Hamburg 2017
Printed in Germany

ACKNOWLEDGMENTS

It is a great opportunity to express our sincere thanks to all who have contributed to do this work through their support, encouragement and guidance.

ABSTRACT

The Floating Point Multiplier is a wide variety for increasing accuracy, high speed and high performance in reducing delay, area and power consumption. The floating point is used for algorithms of Digital Signal Processing and Graphics. Many floating point multipliers are used to reduce the area that perform in both the single precision and the double precision in multiplication, addition and subtraction.

The scientific notations sign bit, mantissa and exponent are used. The real numbers are divided into two, fixed component of significant range (lack of dynamic range) and exponential component in floating point (largest dynamic range). Converting decimal to floating point and normalize the exponent part and rounding operation for reducing latency. The mantissa of two values are multiplied and adding the exponent part. The sign result with exclusive-or are obtained. The final result of shift and add floating point multiplier is compared with booth multiplication. From the synthesis it is observed that shift and add floating point multiplier performs 20% better in different parameters such as delay, area and power is verified by Verilog Hardware Description Language. These results are also verified in Cadence EDA tool.

TABLE OF CONTENTS

CHAPTER NO	TITLE	PAGE NO
	ABSTRACT	ii
	LIST OF TABLES	vii
	LIST OF FIGURES	viii
	LIST OF ABBREVIATIONS	x
1	**INTRODUCTION**	1
	1.1 Format Parameters	1
	1.2 Data formats for single and double precision	2
	1.3 Representation of the floating point	2
	1.3.1 Denormalized	3
	1.3.2 Over flow	3
	1.3.3 Under flow	3
	1.3.4 Infinity	3
	1.3.5 Not a Number	3
2	**LITERATURE SURVEY**	4
	2.1 Floating point multiplier using Vedic mathematics	4

		2.2 Floating point operation in fast fourier transform	4
		2.3 Multiplication using carry save multiplier	5
		2.4 Parallel implementation of floating point	5
		2.5 Normalization of floating point	6
		2.6 Configurable booth multiplier	6
		2.7 Implementation on FPGA	7
		2.8 Different multipliers	7
		2.9 Double precision floating point	8
3		**FLOATING POINT MULTIPLICATION OPERATION**	9
		3.1 Fraction	9
		3.2 Representation of floating point multiplication	10
		3.3 Floating point algorithm	11
		3.3.1 Convert 2.625 to floating point format	11
		3.3.2 Adding an exponent part to binary number	11
		3.3.3 Normalization	12
		3.3.4 Mantissa	12

	3.4 Multiplication operation	13
	3.5 Multiplication of mantissa	14
	3.6 Adding the exponents	15
	3.7 Calculation	16
4	**SIMULATION IN ISE**	**17**
	4.1 VHDL of adding the exponents	18
	4.2 Multiplication	19
	4.3 Multiplication of mantissa	19
5	**SCHEMATIC GENERATED IN ISE**	**21**
	5.1 Actual Schematic	21
	5.2 Schematic for Mantissa	21
	5.3 Magnified image of Mantissa	22
	5.4 Schematic for Exponent	22
	5.5 Schematic for Sign	23
6	**SIMULATION AND CALCULATION OF POWER IN ISE DESIGN**	**24**
	6.1 Synthesis power	24
	6.1.1 Device Static Power	25
	6.1.2 Design Power	25
	6.1.3 Power-On Current	25
	6.1.4 Total On-Chip Power	26

	6.1.5 Off-Chip Power	26
	6.2 Simulation for top module	26
7	**BOOTH MULTIPLICATION OPERATION**	27
	7.1 Algorithm for booth multiplication	27
	7.2 Multiplication operation	28
8	**SIMULATION**	30
9	**SCHEMATIC GENERATED**	32
	9.1 Actual Schematic	32
	9.2 Schematic for booth multiplier	33
	9.3 Schematic for Exponent	33
	9.4 Schematic for Sign	34
10	**SIMULATION AND POWER**	35
	10.1 Comparison of floating point multiplier and booth multiplier	36
11	**IMPLEMENTATION IN FPGA KIT**	37
	11.1 RTL Schematic	37
	11.2 Process	38
	11.3 The input to control FPGA kit	38
12	**CONCLUSION**	40
	REFERENCES	41
	APPENDIX	43

LIST OF TABLES

TABLE NO	TITLE	PAGE NO
1.1	Data formats for precision	2
3.1	Basic formats for IEEE standards	9
3.2	Example result in 8bit and 32bit	13
3.3	Multiplication operation	14
3.4	Multiplication of Mantissa	14
3.5	Result in 32 bit	15
3.6	Final results of sign, exponent and mantissa	16
6.1	Power analysis	25
10.1	Comparison of floating point array and booth multiplier	36

LIST OF FIGURES

FIG NO	TITLE	PAGE NO
1.1	Floating point representation	2
3.1	Block diagram of floating point multiplier	10
4.1	Simulation of adding exponents	18
4.2	Multiplication	19
4.3	Simulation of Multiplying Mantissa	20
5.1	Actual Schematic	21
5.2	Schematic for Mantissa	21
5.3	Magnified image of Mantissa	22
5.4	Schematic for Exponent	22
5.5	Schematic for Sign	23
6.1	Synthesis power	24
6.2	Simulation for top module	26
7.1	Block diagram of Booth multiplier	27
8.1	Simulation of booth multiplier	31
9.1	Actual Schematic	32
9.2	Schematic for Booth multiplier	33
9.3	Schematic for Exponent	33

9.4	Schematic for Sign	34
10.1	X power analyzer for Booth multiplication	35
11.1	RTL Schematic	37
11.2	The input to control FPGA kit	38
11.3	The output to control FPGA kit	39

LIST OF ABBREVIATIONS

LZA	- Leading Zero Anticipation
Model SIM	- Model Simulation
FPGA	- Field Programmable Gate Array
IEEE	- Institute of Electrical and Electronic Engineer
ASIC	- Application-Specific Integrated Circuit
RTL	- Register Transfer Level
VHDL	- Very high speed integrated circuit Hardware Description Language
VHSIC	- Very High Speed Integrated Circuit
ISE	- Integrated Software Environment

CHAPTER 1
INTRODUCTION

The demand of floating point multiplier is more in Three-Dimensional (3D) array and also used in graphics and image processing. Fast Fourier Transform (FFT), Discrete Cosine Transform (DCT) and Butterfly operations are needed of floating point numbers [1]. Due to output data size is twice larger than the input data size so complexity, area and time are consumed by the multipliers. The best design challenge to get high speed working is in Field Programmable Gate Array (FPGA). The floating point shows the base, the location, the precision and it normalized or not. There are many models for multiplication floating point. Precision is the main role in floating point.

We deal with both single and double precision floating point. The main significand of floating point number are (Sign bit * Mantissa * Base ^{Exponent}). The single precision has 24-bits which contain 0 to 31, left to right and double precision have 64-bits which contain 0 to 63, left to right [2]. The difference of these two precision is data, the double precision has twice the data of RAM, Cache and Band Width and reduce the performance. The result of sign bit by XOR and carry save adder used for two exponent components.

1.1   Format Parameters

The implementation of hardware and software has basic IEEE format. In the standard IEEE format the floating points are in binary number. The binary floating point numbers are single precision and double precision. The single precision contains 32 bits and the precision which adds the fraction and hidden bits 23+1, exponent bit 8 is used. The maximum and the minimum values from +127 to -126. The logic utilization of double precision is more by 49% when compared to single precision format. However double precision fused unit exhibits high level of precision when compared to single precision representation.

1.2 Data formats for single and double precision

Table1.1: Data formats for Precision

PRECISION	SIGN BIT	BIASED EXPONENT	UNSIGNED FRACTION
Single	S	8 bit – E	23 – bits P
Double	S	11 bit – E	53– bits P

For the double precision, contain 64 bits and the precision which has 52+1, exponent bit is 11 is used. The maximum and the minimum values from -126 to -1022. For quadruple precision, hidden bits are 112+ 1 and the maximum and the minimum value from -16382 to + 16383 [3].

1.3 Representation of the floating point

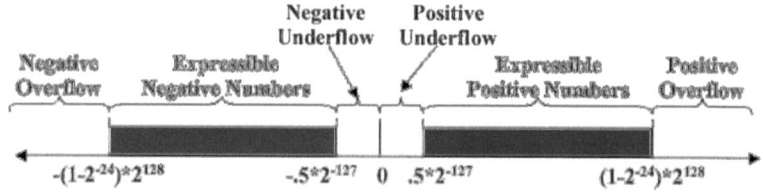

Figure1.1: Floating point representation

Source: https://www.cise.ufl.edu/~mssz/CompOrg/CDA-arith.html

1.3.1 Denormalized

Exponent part contain zeros and fraction or significand contain non-zeros denormalized is taken. Denormalized occur in zeros and lower normalized range [3]. Zero is a special value for exponent field all zeros and fraction zeroes.

1.3.2 Overflow

Overflow occur limited range in smallest value and higher range in highest value. It indicate the range when reach extreme value. It doesn't show the indication when one operand is infinity. It must have the exact range [4]. When the result reaches extreme range, bias should adjust and a NaN is delivered instead.

1.3.3 Underflow

Underflow takes place when floating point is smaller than the smallest value. It may be negative or positive exponent from -128 to 127, when lesser than -128 underflow occur. The result may be zero or denormal [4]. There is loss of accuracy after the denormalized numbers. Under flow adjust the result from overflow delivery.

1.3.4 Infinity

The value of -infinity and +infinity used in exponent 0s and 1s. Sign bit for positive 0 and negative 1 are used. It denotes infinity as special value for operations to continue past overflow situations. It used undefined operations [5].

1.3.5 Not a Number

It is an invalid value when does not show the real number representation. The exponent has 1s and the fraction has non-zeroes are taken in NaNs [6].

CHAPTER 2

LITERATURE SURVEY

2.1 FLOATING POINT MULTIPLIER USING VEDIC MATHEMATICS

Radhika Jumde, IJE [1], The research on the floating point multiplier "Design of 32 bit single precision floating point multiplier using vedic mathematics" is described. To improve delay a new algorithm called Urdhva-Triyakbhyam will be design for the multiplier design. By using this approach number of components will be decreased and complexity of hardware circuit will also be decrease. In this project, Vedic multiplication technique will be used to design IEEE 754 floating point multiplier. The sign bit of the result is calculated using one XOR gate and a Carry Save Adder is used for adding the two biased Exponents. The underflow and overflow cases are handled. The inputs to multiplier will be design, synthesize and stimulate in VHDL using Xilinx ISE tool. This result shows the high speed of multiplication with carry save adder.

2.2 FLOATING POINT OPERATION IN FAST FOURIER TRANSFORM

E. E. Swartzlander, Jr. and H. H. Saleh, 2012 [2], "FFT implementation with fused floating point operation", IEEE Trans. This describes a floating-point fused dot-product unit is presented that performs single-precision floating-point multiplication and addition operations on two pairs of data in a time that is only 150% the time required for a conventional floating-point multiplication. When placed and routed in a 45nm process, the fused dot-product unit occupied about 70% of the area needed to implement a parallel dot-product unit using conventional floating-point adders and multipliers. The speed of the fused dot-product is 27% faster than the speed of the conventional parallel approach. The fused two-term dot-product multiplies two sets of operands and adds the products as a single operation. The two products do not need to be rounded (only the sum is normalized and rounded) which reduces the delay.

2.3 MULTIPLICATION USING CARRY SAVE MULTIPLIER

U. V. Chaudhari, Prof. A. P. Dhande, IJSER [3]. "Design and simulation of binary floating point multiplier using VHDL". This paper shows the possible ways to represent real numbers in binary format floating point numbers are represents two floating point formats, Binary interchange format and Decimal interchange format. To improve speed multiplication of mantissa is done using specific multiplier replacing Carry Save Multiplier. To give more precision, rounding is not implemented for mantissa multiplication. The binary floating point multiplier is plane to do implement using VHDL and it is simulated and synthesized by using Modal Sim and Xilinx ISE software. This multiplier doesn't implement rounding and presents the significand multiplication result as is (48 bits), this gives better precision if the whole 48 bits are utilized in another unit.

2.4 PARALLEL IMPLEMENTATION OF FLOATING POINT

S.Kishore, S.P.Prakash, IJIRSET [4], "The paper floating point fused add-subtract and fused dot-product units" is presented that performs simultaneous floating-point add and multiplication operations. It takes to perform a major part of single addition, subtraction and dot-product using parallel implementation. This unit uses the single-precision format and supports all rounding modes. The fused add-subtract unit is only about 56% larger than a conventional floating-point multiplier, and consumes 50% more power than the conventional floating-point adder. The speed of the fused dot-product is about 27% faster than the conventional parallel approach. This will combine to use for FFT algorithms mainly. The simulation results are obtained using Xilinx 14.3 EDA tool. The proposed system reduces the shift amount and normalization is applied to reduce the size of significand addition and LZA reduces the reduction tree.

2.5 NORMALIZATION OF FLOATING POINT

V.Narasimha Rao, V.Swathi [5], "Normalization on floating point multiplication using Verilog HDL". This paper describes an efficient implementation of an IEEE 754 single precision floating point multiplier targeted for FPGA. VHDL is used to implement a technology-independent pipelined design. The multiplier implementation handles the overflow and underflow cases. Rounding is not implemented to give more precision when using the multiplier in a multiply and Accumulate (MAC) unit. A latency optimized floating point unit using the primitives of Xilinx Virtex II FPGA was implemented with a latency of 4 clock cycles. The multiplier reached a maximum clock frequency of 100 MHz.

2.6 CONFIGURABLE BOOTH MULTIPLIER

K. Sreenath, K. Shashidhar [6], "The design of low power and high speed configurable booth multiplier". This design of low power and high speed configurable Booth multiplier (CBM) that supports single 16*16 multiplication, single 8*8 multiplication, and twin parallel 8*8 multiplication operations. To efficiently reduce power consumption, a dynamic-range detector is developed to dynamically detect the effective dynamic ranges of two input operands. The detection result is used not only to pick the operand with smaller dynamic range for Booth encoding but also deactivate the redundant switching activities in ineffective ranges as much as possible. Moreover, the output product of the proposed multiplier can be truncated further which results in decrease power consumption by sacrificing a bit of output precision. To obtain accurate results, some additional components, including a sign-bit generator, a modified error compensation circuit, dadda compressor, carry look ahead adder are also developed. The results show that the proposed multiplier is complex than non-CBMs, but significant power savings can be achieved. Furthermore, the paper shows proposed multiplier maintains acceptable output accuracy where truncation is performed.

2.7 IMPLEMENTATION ON FPGA

K.Kishore Shinde, A.K Kureshi [7], "Hardware implementation of configurable booth multiplier on FPGA". This paper presents an efficient implementation of a high performance configurable Radix-4 Booth multiplier with 3:2 compressors for both signed and unsigned 32 bit numbers multiplication & the floating point arithmetic. Multiplication operation is a mostly used in many scientific and signal processing applications. Thus it provides a flexible arithmetic capacity and a better output precision and high speed, minimum area consumption. The design also dynamically disables the switching operation of the non effective input ranges. Thus the ineffective circuits can be efficiently deactivated, thereby reducing power consumption and increasing the speed of operation. The proposed design of multiplier out performs the conventional multiplier in terms of area and speed efficiencies. The multiplier designs have been implemented on FPGA Spartan6 XC6SLX9 platform. VHDL code is written to generate the required hardware and to produce the partial product for proposed booth multiplier. After the successful compilation the RTL view generated.

2.8 DIFFERENT MULTIPLIERS

Soniya, Suresh kumar [8], "A Review of Different Type of Multipliers and Multiplier-Accumulator Unit". High speed and low power MAC unit is utmost requirement of today's VLSI systems and digital signal processing applications like FFT, Finite impulse response filters, convolution etc. In this paper, They have discussed different types of multipliers like booth multiplier, combinational multiplier, Wallace tree multiplier, array multiplier and sequential multiplier. Each multiplier has its own advantages and disadvantages. Different types of techniques are presented for improving the speed and low power consumption like pipelined booth multiplication technique in which pipelining is used in booth multiplier to reduce the delay of each stage. In SPST technique useless portion of the data is removed for reducing the power consumption. And block enabling technique is also used for low power consumption.

2.9 DOUBLE PRECISION FLOATIG POINT

Gargi s. Rewatkar, "Implementation of double precision floating point multiplier in VHDL". Floating point numbers are one possible way of representing real numbers in binary format; the IEEE 754 standard presents two different floating point formats, Binary interchange format and Decimal interchange format. Multiplying floating point numbers is a critical requirement for DSP applications involving large dynamic range. Floating-point implementation on FPGAs has been the interest of many researchers. FPGAs are increasingly being used in the high performance and scientific computing community to implement floating-point based hardware accelerators. FPGAs are generally slower than their application specific integrated circuit (ASIC) counterparts and draw more power. However, they have several advantages such as a shorter time to market, ability to re-program in the field to fix bugs, and lower nonrecurring engineering cost costs. The development of these designs is made on regular FPGAs and then migrated into a fixed version that more resembles an ASIC.

CHAPTER 3

FLOATING POINT MULTIPLICATION OPERATION

3.1 FRACTION

The real numbers are divided into two parts. They are integer and fractional part. An integer part on the left to the radix point with using positive powers and a fraction part on the right to radix point with using negative powers. The fraction part does not give exact representation, so it divides as fixed and floating point numbers. The fixed component of significant range (lack of dynamic range) and exponential component in floating point (largest dynamic range).

According to IEEE 754 2008 supports the floating point multiplier which has efficient carry saver. For the high performance of multiplier, pipe lining stages are used to increase operating frequency multiplier [7].

Table3.1 Basic formats for IEEE standards

Name	Common name	Base	Digits	E_{min}	E_{max}
binary 32	Single precision	2	23+1	-126	+127
binary 64	Double precision	2	52+1	-1022	+1023
binary 128	Quadruple precision	2	112+1	-16382	+16383
decimal 64	-	10	16	-383	+384
decimal 128	-	10	34	-6143	+6144

Here two approaches are seen for dot product unit, floating point adder and multiplier. In general, a floating-point number consists of three main parts: sign (S), mantissa (M) and exponent (E). The precision is the important in floating point numbers. They are single precision and double precision. The double precision has the output data twice the input data when compare to the single precision.

3.2 REPRESENTATION OF FLOATING POINT MULTIPLICATION

The IEEE standard specifies basic and extended floating-point number formats, arithmetic operations, conversions between various number formats, rounding algorithms, and floating-point exceptions. All the basic formats may be available in both hardware and software implementations. The fused dot product unit derived from floating point add sub unit. The floating point multiplier block diagram is shown. The two exponents are biased and at same time mantissa or significand combined to get parallel precision. The combined results are normalized. They done separately and multiplexers choose add and sub with XOR process.

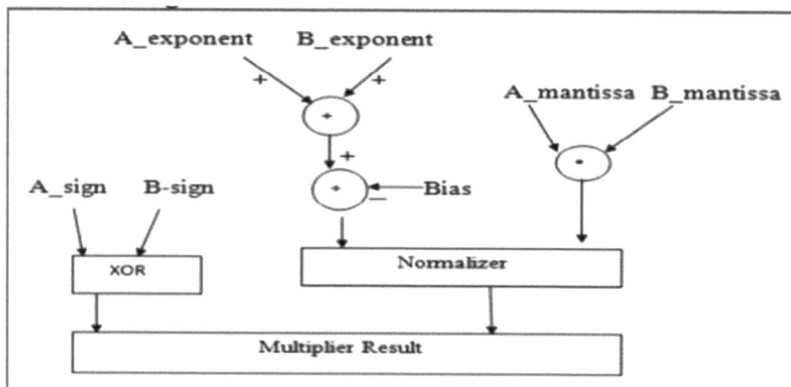

Figure3.1: Block diagram of floating point multiplier
Source: http://www.ijser.org/paper/Design-And-Simulation-Of-Binary-Floating-Point-Multiplier-Using-VHDL.html

Converting decimal number to the floating point number, block diagram for floating point multiplier is shown. The two input values of exponents are added and biased. The two input mantissa values are multiplied. The two sign bits are process the XOR. Then the results of mantissa, biasing and XOR are results the floating point multiplication.

3.3 FLOATING POINT ALGORITHM

Convert the value to binary, take fraction for separating the integral value and fractional value. This fractional part is converting by multiplication [8]. Multiply by 2 repeatedly and harvest each one bit. It shows the decimal value to floating point.

For Example:

3.3.1 Convert 2.625 to floating Point format

The integral part is $2_{10} = \mathbf{10_2}$

The fractional parts are

0.625*2= 1.25 --**1**

0.25 *2= 0.5 -- **0**

0.5 *2= 1.0 -- **1**

For $0.625_{10} = \mathbf{0.101_2}$ $2.625_{10} = \mathbf{10.101_2}$

3.3.2 Adding an exponent part to binary number

One possibility for handling numbers with fractional parts is to add bits after the decimal point: The first bit after the decimal point is the halves place, the next bit the quarters place, the next bit the eighths place and to split the bits of the representation between the places to the left of the decimal point and places to the right of the decimal point. For example, a 32-bit fixed-point representation might allocate 24 bits for the integer part and 8 bits for the fractional part. The Product of append and 2 power exponent in the end of binary numbers.

$10.101_2 = 10.101_2 * 2^0$

3.3.3 Normalization

The mantissa of a floating point number represents an implicit fraction whose denominator is the base raised to the power of the precision. Since the largest represent mantissa is one less than this denominator (base raise to the power of the precision), the value of the fraction is always strictly less than 1. The mathematical value of a floating point number is then the product of this fraction, the sign, and the base raised to the exponent. If the number is not normalized, then you can subtract 1 from the exponent while multiplying the mantissa by the base, and get another floating point number with the same value. Normalization consists of doing this repeatedly until the number is normalized. Two distinct normalized floating point numbers cannot be equal in value.

The value doesn't change when exponent adjust to one bit left in binary number. $10.101_2 * 2^0 = 1.0101_2 * 2^1$

3.3.4 Mantissa

Mantissa or significand is next to leading number which filled with zeroes on the right [9]. When working in binary, the significand is characterized by its width in bits. Because the most significant bit is always 1 for a normalized number, this bit is not typically stored and is called the hidden bit. Depending on the context, the hidden bit may or may not be counted towards the width of the significand. For example, the same double precision format is commonly described as having either a 53-bit significand, including the hidden bit, or a 52-bit significand, not including the hidden bit. The notion of a hidden bit only applies to binary representations.   Mantissa – **0101**

Now add the bias to exponent of 2 in exponent field, for all biasing $2^{k-1} - 1$, k=number of bits in exponent field. For example 8– bit format $2^{3-1} -1 = 3$

32 –bit format $2^{8-1} -1 = 127$.

Here the exponent power is 1, so $1 + 3 = 4 = 100_2$

1+127= 128=1000000

The sign bit of negative is 1 and positive is 0 of given number. For 8 bit and 32 bit

Table3.2: Example result in 8bit and 32bit

SIGN BIT	MANTISSA	EXPONENT
0	100	0101
0	10000000	01010000000000000000000

3.4   MULTIPLICATION OPERATION

Multiplying the two input values after the normalization. The sign, exponent and mantissa are taken separately. The multiplication must take account of the integer part, implicit in normalization. The number of bits of the result is twice the size of the operands (48 bits).

Biasing is done because of exponent have to be signed values in order to represent both tiny and huge values. The exponent is biased before stored, by adjusting its value to put it with in an unsigned range suitable range for comparison.

Take 2 floating point number A=3 and B=12

Their binary values are

A= 00000011

B= 00001100

The normalization of

A= $1.1*2^{-3}$ B=$1.1*2^{3}$

Table3.3: Multiplication operation

SIGN BIT	MANTISSA	EXPONENT
0	01111100	10000000000000000000000
0	10000010	10000000000000000000000

3.5   MULTIPLICATION OF MANTISSA

For normalization [10] adding 1 to the most significant bit is useful,

Table 3.4: Multiplication of mantissa

A	10010000000000000000000
B	10011000000000000000000
A*B	01010101100000000000000

We must extract the mantissa, adding 1 as most significant bit, for normalization the result of the multiplication only the most significant bits are useful: after normalization (elimination of the most significant 1), we get the 23-bit mantissa of the result.

This normalization can lead to a correction of the result's exponent. The multiplication is more complex for floating point and integer. The multiplication of mantissa is given in the table. The result is in 32 bit: 24_H

Table 3.5: Result in 32 bit

SIGN BIT	MANTISSA	EXPONENT
01	00100100	000000000000000000000000

3.6 ADDING THE EXPONENTS

Exponent of the result is equal to the sum of the operands exponents. A 1 can be added if needed by the normalization of the mantissa multiplication. For mantissa multiplication remove bias in two operands and add again the bias [10].

E result= (Ea-127) + (Eb-127) +127

E r= Ea+ Eb-127

then E r= 00000011

      + 00001100

      -01111111

      = 1110010

3.7 CALCULATION

The setting of the 3 intermediate results (sign, exponent and mantissa) gives us the final result of our multiplication. The sign of the result (Sr) is given by the exclusive-or of the operands signs (Sa and Sb).

The Sign result -Sr by EXOR of two operands

Sa= 0 and Sb= 0 is

Sr= Sa⊕Sb

Sr= 0⊕ 0 = 1

Table 3.6: Final results of sign, exponent and mantissa

SIGN BIT	MANTISSA	EXPONENT
1	1110010	01010101100000000000000

$A*B = 3* 12 = 1.00100*2^{134-127} = 10101011.0 = 171_{10}$

The calculation for the floating point multiplier is obtained and sign bit is 1 and mantissa 8 bit value is 1110010 and exponent of 23 bit is 01010101100000000000000.

CHAPTER 4

SIMULATION IN ISE

The fused floating point dot product multipliers of 32 bit in double precision reduce the delay and silicon area. It increases the speed faster than single precision floating point. It also decreases 40% of the worst errors. The discrete design is the worst case for reducing the area and consumption [11]. Floating point calculation design has been simulated in ISE design, create design in work library as 45nm CMOS study cell and compile the design by go to simulation and start simulate to run, before that source code in VHDL [12].

The VHSIC (very high speed integrated circuits) Hardware Description Language (VHDL) was first proposed in 1981. The development of VHDL was originated by IBM, Texas Instruments, and Inter-metrics in 1983. The result, contributed by many participating EDA (Electronic Design Automation) groups, was adopted as the IEEE1076 standard in December 1987.VHDL is a high level language and used for digital circuits and systems. This representation allows easy description and direct functionality through simulation. The system verilog input generates stimulus output by the components of entity and architecture, connecting the configuration simulation [13]. It is easy and similar to C programming and also used in Engineering Design Automation Tool [14]. We can see that the addition of the multiplier and the aligned by carry save adder. The bit calculation which is needed for rounding and the anticipation of the leading zeros (LZA) which is needed for normalization performed parallel with carry save adder.

ISE simulator is a hardware description language (HDL) event-driven simulator that supports behavioral and timing simulation for single language and mixed language design. The simulation in ISE is a feature-rich, mixed-language that supports the VHDL. The compilation enabled to improve compile time significantly. For Synthesis and analysis of HDL, Xilinx ISE development and high-level synthesis are used. Delivering pre-compiled libraries with ISE design simulators.

4.1 VHDL SIMULATIONS FOR ADDING THE EXPONENTS

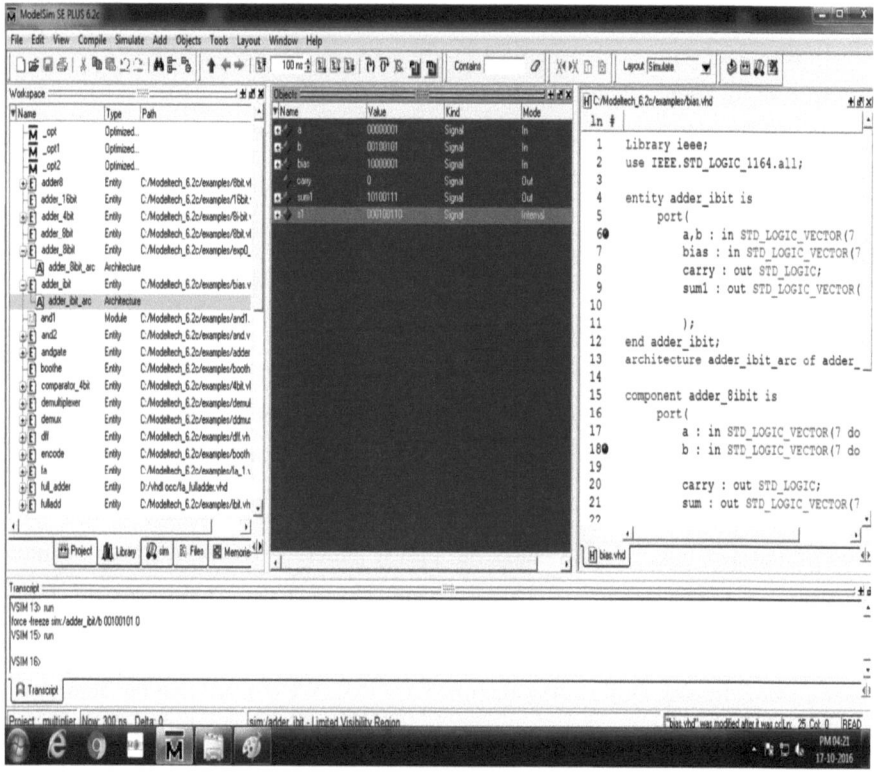

Figure4.1: Simulation of adding exponents

The two inputs values are converted to binary values and they are separated to sign, exponents and mantissa. Here, the two exponents are added and simulated with sum, carry and bias value. The simulation using Model SIM, the results are executed with the programs.

4.2 MULTIPLICATION

The two inputs values of binary numbers are multiplied and the signal is given for fast execution. The results are executed; the product value in 23 bit is shown by source program.

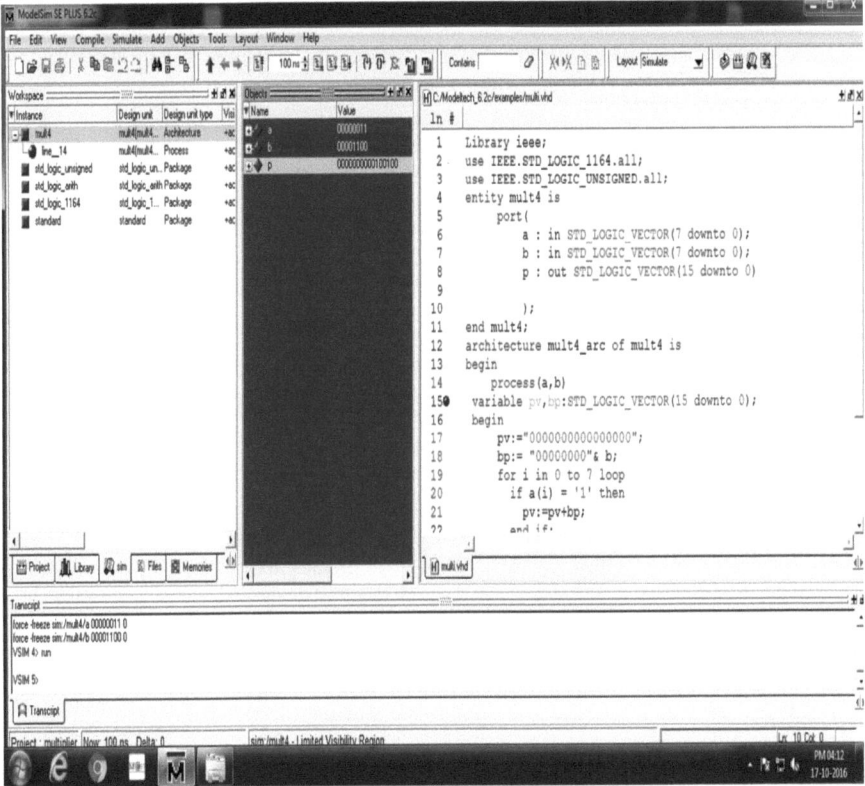

Figure4.2: Multiplication

4.3 MULTIPLICATION OF MANTISSA

The two inputs values in floating point numbers of mantissa is multiplied and executed the results are shown. By giving the signal value, two 8 bits are multiplied and resulted in 16 bit conversion. The result of the significand multiplication (intermediate product) must be normalized to have a leading 1 just to the left of the decimal point (i.e. in the bit 23 in the intermediate product). Since the inputs are normalized numbers then the intermediate product has the leading one.

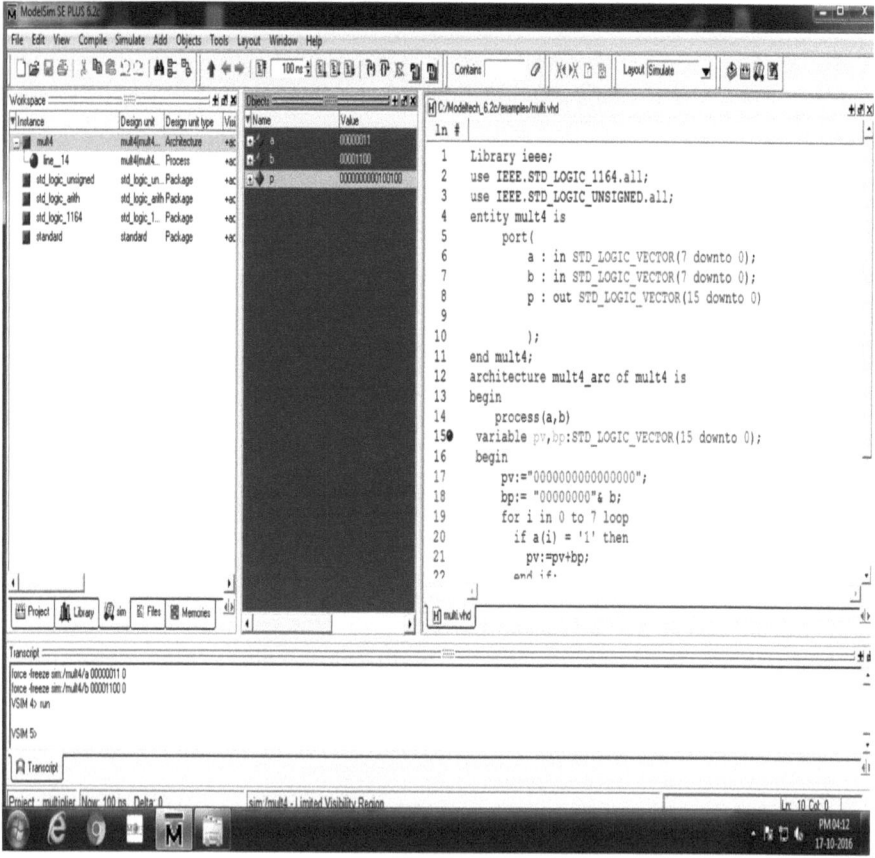

Figure4.3: Simulation of Multiplying Mantissa

CHAPTER 5

SCHEMATIC GENERATED IN ISE

5.1 Actual Schematic

The diagram for sign, exponent and mantissa are executed as top module and the actual schematic diagram is shown.

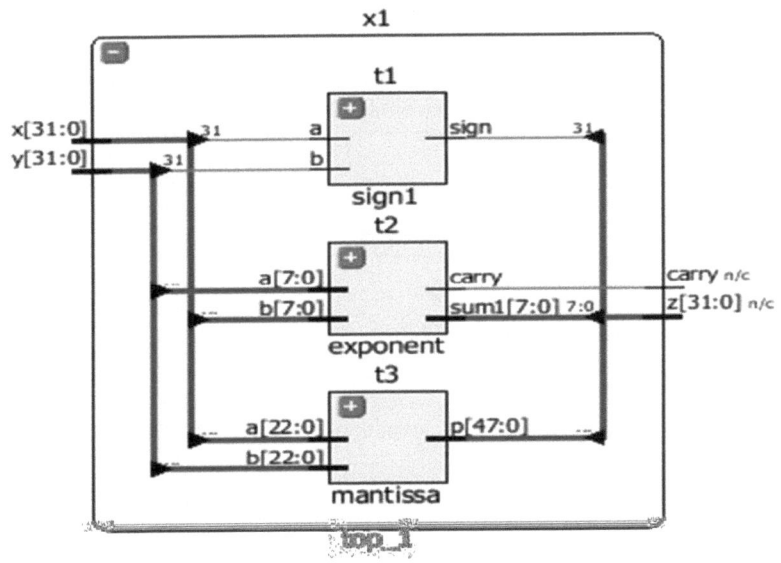

Figure5.1: Actual Schematic

5.2 Schematic for Mantissa

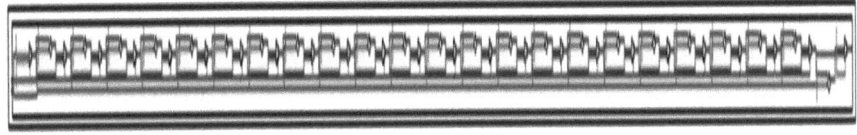

Figure5.2: Schematic for Mantissa

The schematic diagram for mantissa is shown and the magnified image of mantissa is also shown below.

5.3 Magnified image of Mantissa

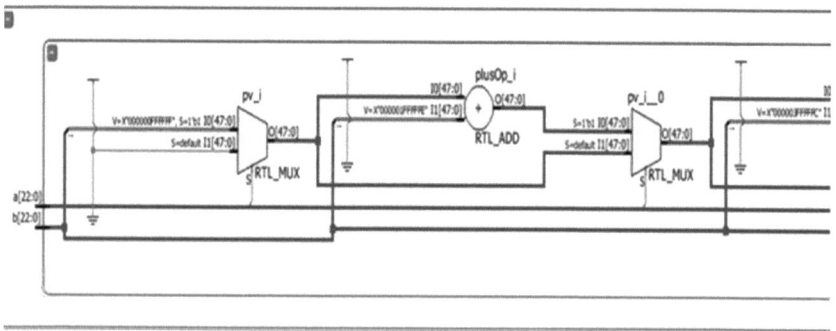

Figure5.3: Magnified image of Mantissa

5.4 Schematic for Exponent

The addition of two 8 bit adder exponent of schematic diagram is shown which executed in ISE Design.

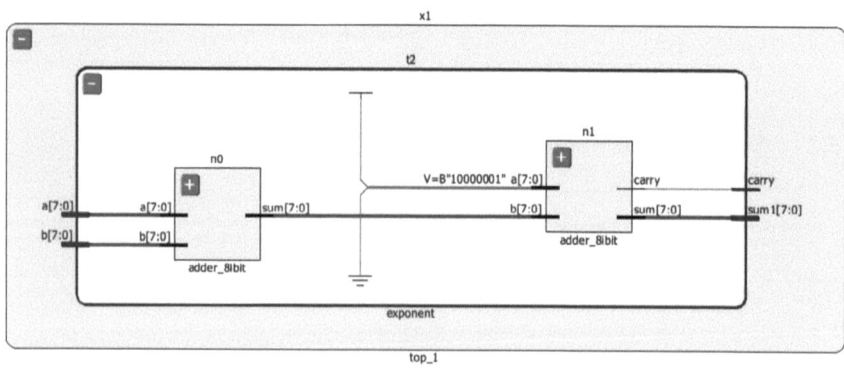

Figure5.4: Schematic for Exponent

5.5 Schematic for Sign

The schematic diagram for sign bit is shown. Here, the RTL_XOR processes are done and results the output. Then top module is executed by combining the three schematic diagrams.

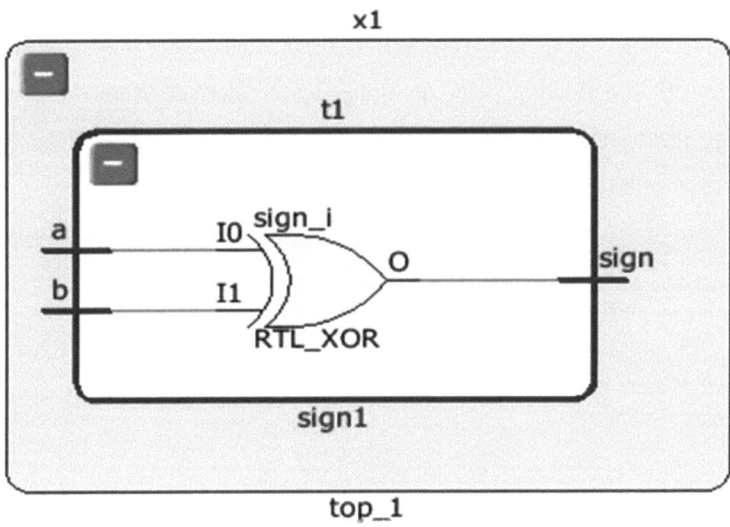

Figure5.5: Schematic for Sign

The Spartan-3E High Volume Starter Kit gives designers instant access to the complete platform capabilities of the Spartan-3E family. The Spartan-3E FPGA Starter Kit is a complete development board solution. Using this result is verified. Complete kit includes board, power supply, evaluation software, resource CD. A few system-level design trade-offs were required in order to provide the Spartan-3E Starter Kit board with the most functionality.

CHAPTER 6

SIMULATION AND CALCULATION OF POWER IN ISE DESIGN

6.1 SYNTHESIS POWER

The output power of the floating point multiplier is executed in ISE design, the resulted power is 42.134 Watts sysnthesised . This power may be change in on-chip power. The floating point multiplier results derived from the combination sign, mantissa and exponent values and mainly from constraints files, simulation files or vector less analysis.

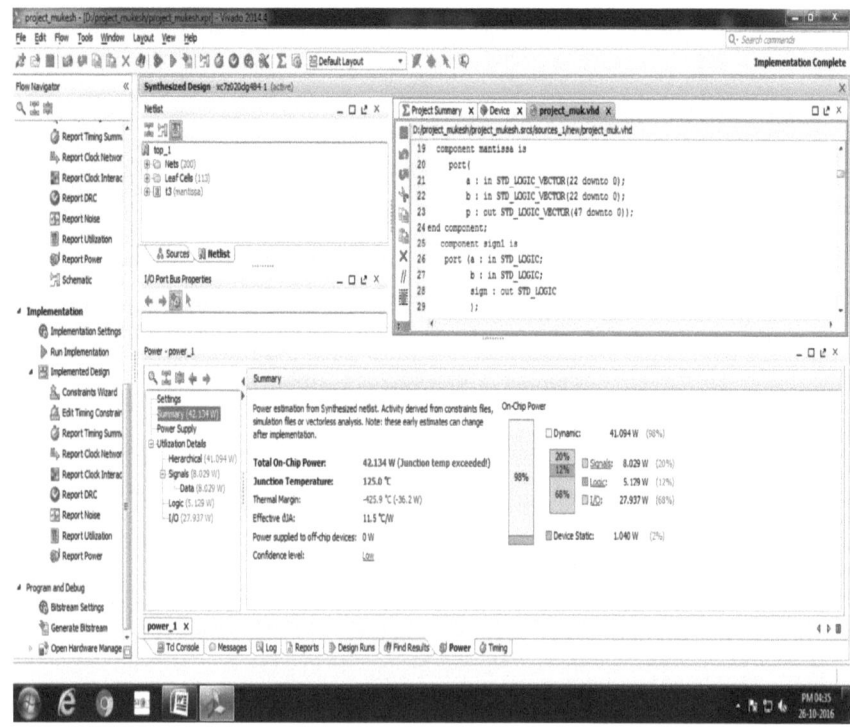

Figure6.1: Synthesis power

6.1.1 Device Static Power

Device static power is the power from transistor leakage on all connected voltage rails and the circuits required for the FPGA to operate normally, post configuration. This is normally measured by programming a blank bit stream into the device. Device static power is a function of process, voltage, and temperature. This represents the steady state, intrinsic leakage in the device.

6.1.2 Design Power

Design power is the power of the user design, due to the input data pattern and the design internal activity. This power is instantaneous and varies at each clock cycle. It depends on voltage levels and logic and routing resources used. This also includes static current from I/O terminations, clock managers, and other circuits that need power when used. It does not include power supplied to off-chip devices.

Table6.1: Power analysis

Total Power	Delay	Power Delay Product
178mW(Typical)	70.646ns	12.574Wns
610mW(Maximum)	70.646ns	43.094Wns

6.1.3 Power-On Current

Power-on current is transient current that occurs when power is first applied to the FPGA. This current varies for each voltage supply and depends on the FPGA construction as well as the ability of the power supply source to ramp up to the nominal voltage. This current also depends on the device's operating conditions, such as temperature and sequencing between the different supplies.

Power-on current is generally lower than operating current due to architectural enhancements as well as adherence to proper power-on sequencing.

6.1.4 Total On-Chip Power

Total on-chip power is the power consumed internally within the FPGA, equal to the sum of device static power and design power. It is also known as thermal power.

6.1.5 Off-Chip Power

Off-chip power is the current that flows from the supply source through the FPGA power pins then out of I/Os and dissipated in external board components. The currents supplied by the FPGA are generally consumed in off-chip components such as I/O terminations, LEDs, or the I/O buffers of other chips, and therefore do not raise the device junction temperature.

6.2 SIMULATION FOR TOP MODULE

The simulation for top module is executed and results are shown. This is the implemented design is derived from schematic diagram of sign, mantissa and exponent. The programmable logic code converted into the schematic and those schematics are form design for floating point and implemented design is shown.

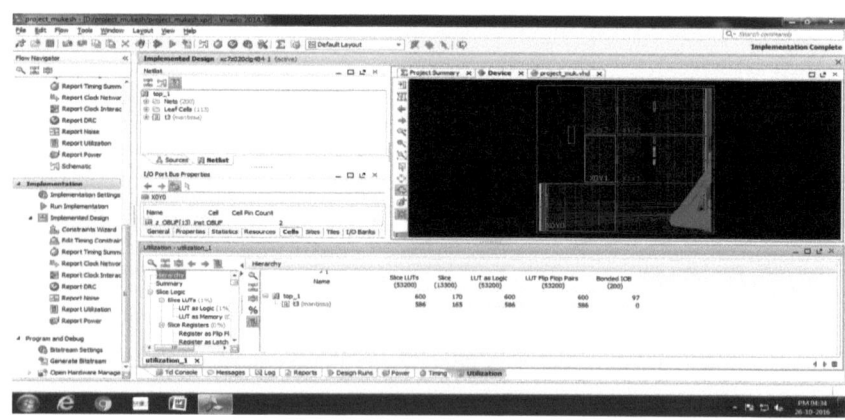

Figure6.2: Simulation for top module

CHAPTER 7

BOOTH MULTIPLICATION OPERATION

When using Booth's Algorithm twice as many bits product as have in original two operands. The left most bit of your operands (both your multiplicand and multiplier) is a sign bit, and cannot be used as part of the value. Decide which operand will be the multiplier and which will be the multiplicand. Convert both operands to two's complement representation using that bits must be at least one more bit than is required for the binary representation of the numerically larger operand [11]. Begin with a product that consists of the multiplier with an additional X leading zero bits.

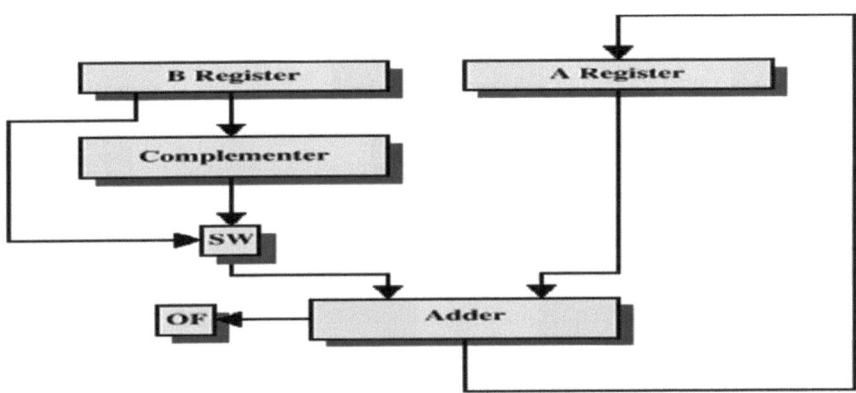

OF = overflow bit
SW = Switch (select addition or subtraction)

Figure7.1: Block diagram of Booth multiplier
Source: http://cnx.org/contents/113da73e-68d1-485c-9a68-61a15b108e4d@1/computer-arithmetic

7.1 ALGORITHM FOR BOOTH MULTIPLICATION

The procedure for booth multiplication is if x is the count of bits of the multiplicand and y is the count of bits of the multiplier. A- Add, S- Subtract and P- Product. In two's complement notation, fill x bit of each line as A is multiplicand,

S is negative of multiplicand and P is zeros. Then fill y bits of each line with A is zeros, S is zeros and P is multiplier and fill last bit of each line with zeros [12]. Booth's algorithm changes the first step of the algorithm—looking at 1 bit of the multiplier and then deciding whether to add the multiplicand—to looking at 2 bits of the multiplier. The new first step, then, has four cases, depending on the values of the 2 bits. Let's assume that the pair of bits examined consists of the current bit and the bit to the right—which was the current bit in the previous step. The second step is still to shift the product right.

If the last two bits of product which possible arithmetic actions are:

00 → no arithmetic operation

01 → add multiplicand to left half of product (P=P+A)

10 → subtract multiplicand from left half of product (P+P+S)

11 → no arithmetic operation

As in the previous algorithm, shift the Product register right 1 bit. Now we are ready to begin the operation. It starts with a 0 for the mythical bit to the right of the right most bit for the first stage. Compares the two algorithms, with Booth's on the right. Note that Booth's operation is now identified according to the values in the 2 bits. By the fourth step, the two algorithms have the same values in the Product register. The one other requirement is that shifting the product right must preserve the sign of the intermediate result, since we are dealing with signed numbers. The solution is to extend the sign when the product is shifted to the right. This shift is called an arithmetic right shift to differentiate it from a logical right shift. Then arithmetic operations are continued for y times, which is multiplier. Add the leading zeros in left half of the product, when possible arithmetic operations are occurred [12].

7.2 MULTIPLICATION OPERATION

Let us take 2 values for multiplying **(-11) * (-4)**

-11 is (multiplier) -4 is (multiplicand)

And we added 5 leading zeros to the **multiplier** to get the **beginning product**.

Initial Product and previous LSB

| A=1 0101 0000 0 | S=0 1011 0000 0 | P=0 0000 1100 0 |

Perform the loop for 4 times because the multiplier or the y term is 4.

1. **P=0 0000 1100 0**

The last 2 bits are **00**, now right shift arithmetic takes place,

P=0 0000 0110 0

2. **P=0 0000 0110 0**

The last 2 bits are **00**, now right shift arithmetic takes place,

P=0 0000 0011 0

3. **P=0 0000 0011 0**

The last 2 bits are **10,** now add the subtract value to product,

P=P+S

SYMBOL	ADDITION
P	0 0000 0011 0
S	0 1011 0000 0

P=0 1011 0011 0.

If the last two bits 01 product should be added to add value and takes right shift arithmetic.

4. **P=0 1011 0011 0**

After adding take right shift arithmetic, **P=0 0101 1001 1.**

The last 2 bits are **11,** now right shift arithmetic takes place, **P=0 0010 1100 1.**

The Product is **0010 1100** now neglect the first and last bit. The result is **44** [13].

CHAPTER 8

SIMULATION

VHDL which stand for "Very high speed integrated circuit Hardware Description Language" is one of the common techniques for the digital system developing process. The technique is done by program using certain software as a platform which also can perform simulation and analysis of the designed system. The designers only need to describe his digital circuit design in textual form which can erase without the effort to alter the hardware.

The programming language is different compared to other programming language such as C++ language. VHDL is more preferred because this technique can reduce cost and time, easy to troubleshoot, portable, a lot of platform software support the VHDL function and high references availability.

VHDL (Very high speed integrated circuit Hardware Description Language) is used as the language for the system. All the process will be running using Xilinx ISE 14.5 software which means the process is simulation only without any hardware implementation.

Booth multiplier is one of the most important parts in the devices which can affect the performance of the devices. So, the high speed and efficient multiplier system is important for the designers of microprocessor, microcontroller and others digital devices. As we know, floating point multiplication operation is not hard to do in decimal number. But, to do the operation in binary number (which used in digital system) is very complex operation. This project is being done to help create a prototype of booth multiplier in digital system design that can operate as multiplier operation that would be implemented into microprocessor, microcontroller and other digital devices.

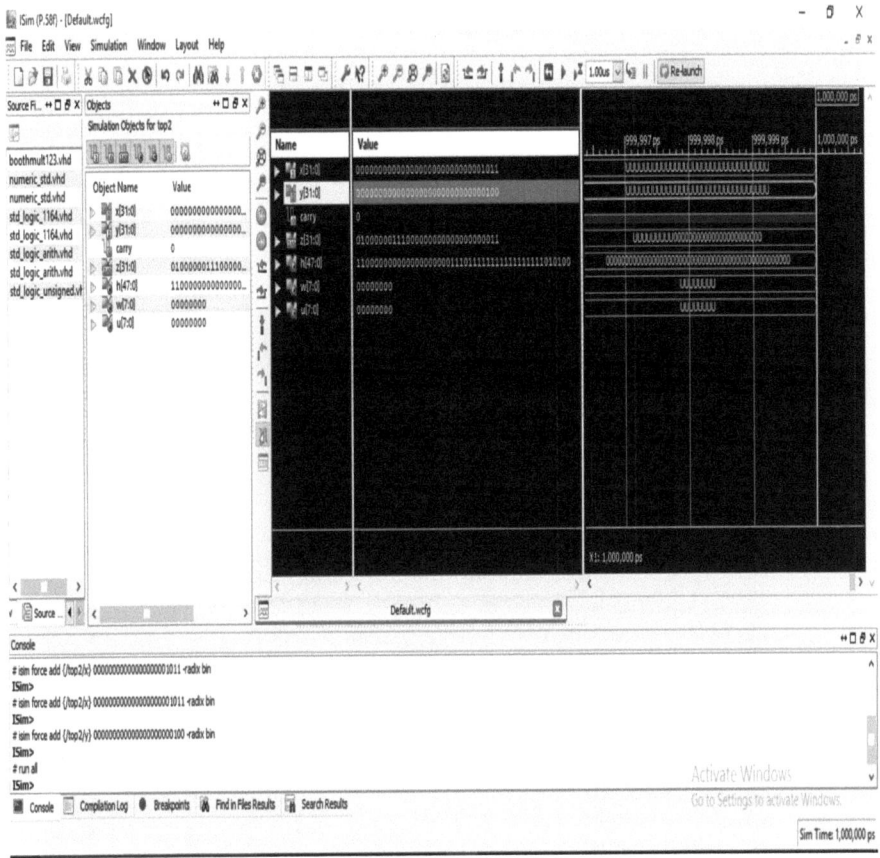

Figure 8.1: Simulation of booth multiplier

The implementation of booth multiplier in VHDL is shown in fig 8.1. It shows the two inputs of binary number carry value and output results with signals. This is the one of the ALU part. Here we do simulation process with the help of the Xilinx ISE 14.5 software. In this simulation process we force value of 'x' as 11 and 'y' as 4 and we get the 'final answer' as 44.

CHAPTER 9

SCHEMATIC GENERATED

9.1 Actual Schematic

The diagram for sign, exponent and booth multiplier is executed as top module and the actual schematic diagram is shown.

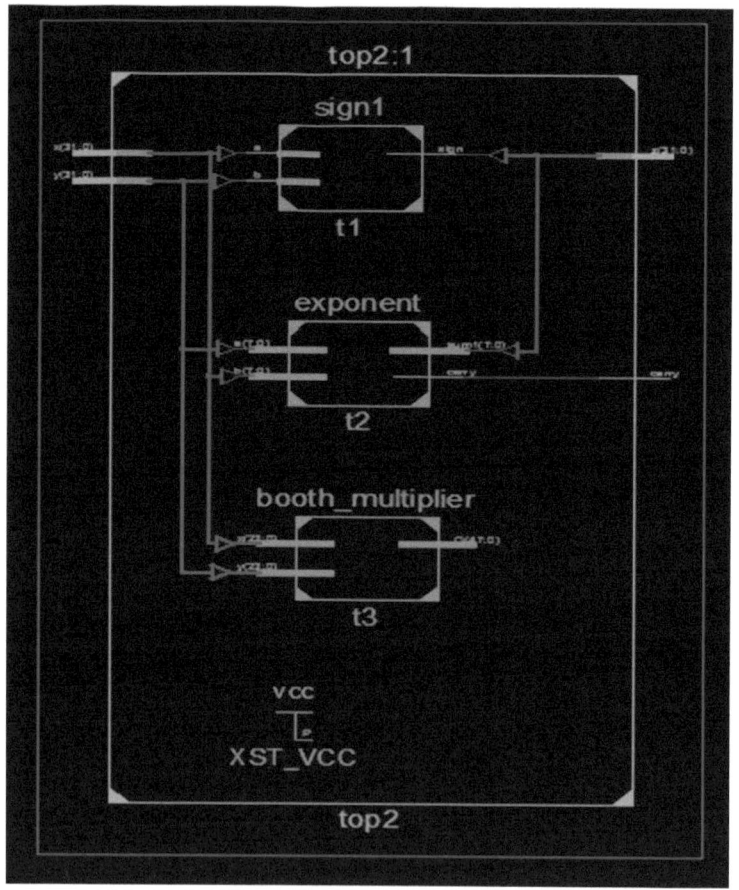

Figure9.1: Actual Schematic

9.2 Schematic for booth multiplier

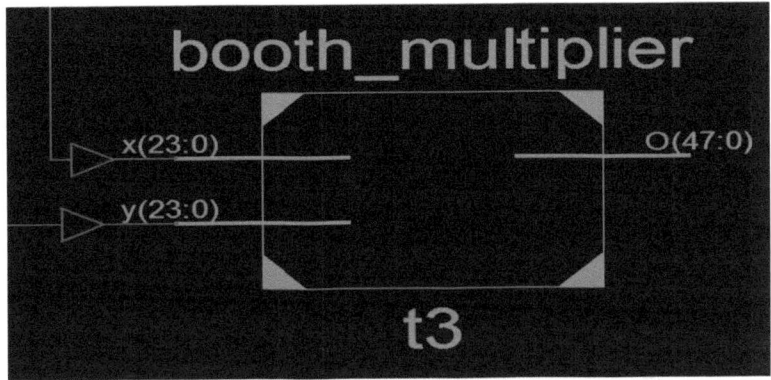

Figure9.2: Schematic for Booth multiplier

9.3 Schematic for Exponent

The addition of two 8 bit adder exponent of schematic diagram is shown which executed in VHDL.

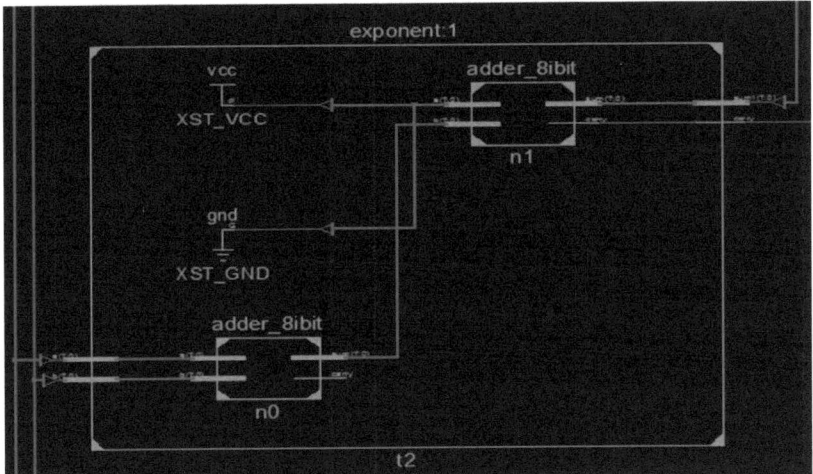

Figure9.3: Schematic for Exponent

9.4 Schematic for Sign

The schematic diagram for sign bit is shown. Here, the RTL_XOR processes are done and results the output. Then top module is executed by combining the three schematic diagrams.

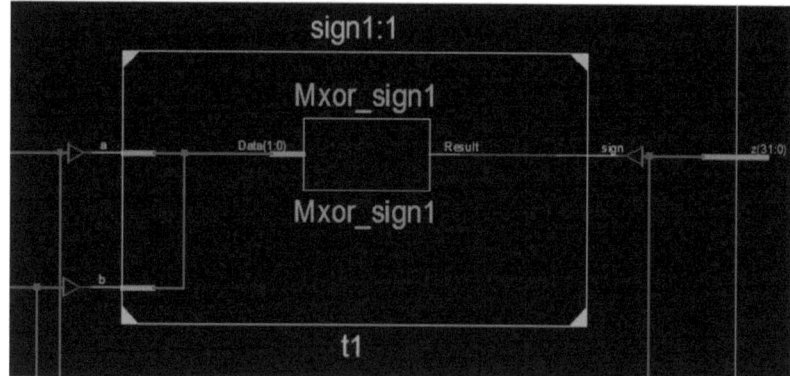

Figure 9.4: Schematic for Sign

CHAPTER 10

SIMULATION AND POWER

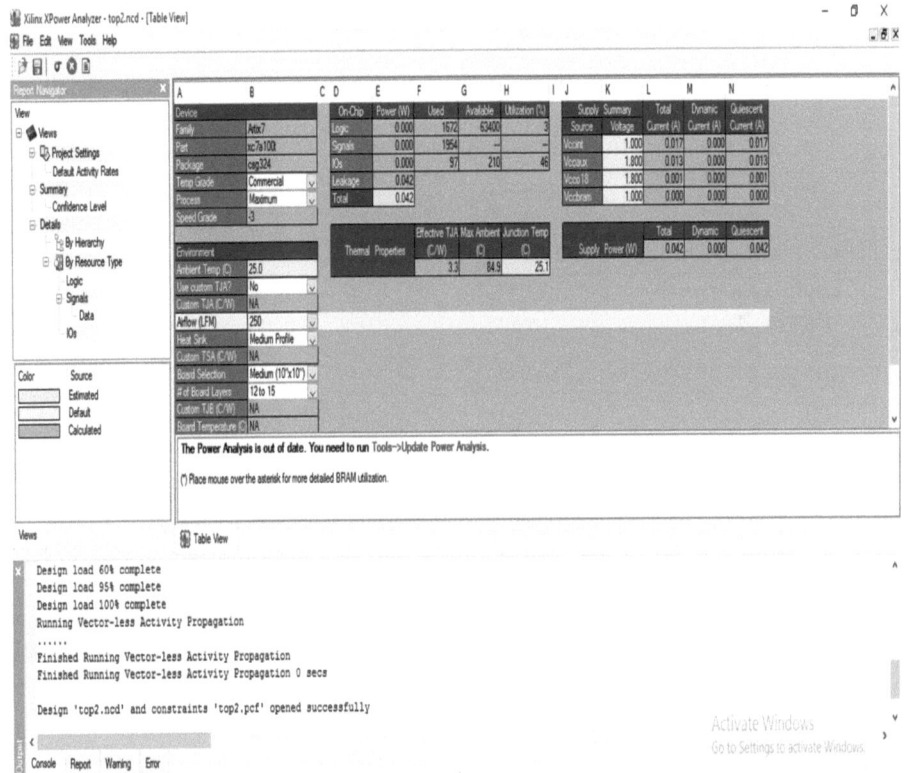

Figure 10.1: X power analyzer for Booth multiplication

As a result of comparison in terms of delay, Booth multiplier has been chosen to proceeds the multiplication of both signed and unsigned numbers. It also reduces the number of partial products. The total power consumptions is 420mWatts in typical and maximum.

10.1 COMPARISON OF FLOATING POINT ARRAY MULTIPLIER AND BOOTH MULTIPLIER

SYNTHESIS	FLOATING POINT SHIFT AND ADD MULTIPLIER	FLOATING POINT BOOTH MULTIPLIER
Area Used	7338.636 (μm^2)	9493.916 (μm^2)
Delay	119-51 (ns)	132-68 (ns)
Leakage Power	156.179 (nw)	172.532 (nw)
Dynamic Power	389775.698 (nw)	579122.171 (nw)
Total Power	398931.878 (nw)	579294.703 (nw)

Table 10.1: Comparison of floating point array and booth multiplier

Thus the comparison of both floating point multiplier and the booth multiplier are executed in VHDL and verified in Cadence digital design which is an automation machine tool. Cadence is an Electronic Design Automation (EDA) environment that allows integrating in a single framework different applications and tools (both proprietary and from other vendors), allowing to support all the stages of IC design and verification from a single environment. These tools are completely general, supporting different fabrication technologies. When a particular technology is selected, a set of configuration and technology-related files are employed for customizing the Cadence environment. This set of files is commonly referred as a design kit. The verified results are showing that the area, power and delay timing are reduced than the booth multiplication. Thus, the floating point array multipliers have greater efficiency and more accuracy than booth multiplier.

CHAPTER 11

IMPLEMENTATION IN FPGA KIT

Field-programmable gate arrays (FPGAs) are reprogrammable silicon chips. It is important to analyze area utilization. Design can be synthesized optimally with respect to area utilization. FPGAs provide hardware-timed speed and reliability. At each level of abstraction, the future integrated circuit is described in HDL, such as behavioral VHDL or synthesized VHDL. In order to simulate and validate the digital circuit's functionality, various test benches are formulated and executed. The flexibility reduced computational time, and possibility to implement complex design in real time.

11.1 RTL SCHEMATIC

The schematic diagram of the synthesized code can be viewed by double clicking view RTL Schematic.

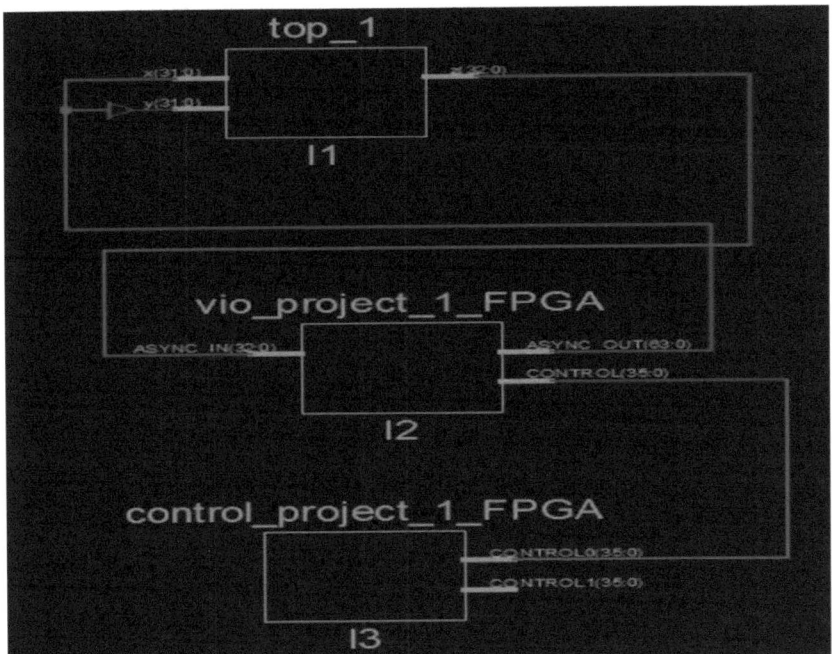

Figure 11.1: RTL Schematic

11.2 PROCESS

1) Use the **Xilinx CORE Generator System** to create an IP core.
2) Connect the IP Core to the **VHDL** source as a component.
3) Synthesize and program the **Spartan 3E Starter Kit** board.
4) Create a new source Implementation Constraints File, and assign the pins as inputs and outputs in order to use **Switch** and **LEDs** of **Spartan-3E FPGA Kit**.

11.3 THE INPUT TO CONTROL FPGA KIT

After the connection and synthesis, the Chip-Scope Analyzer assigns the input and output.

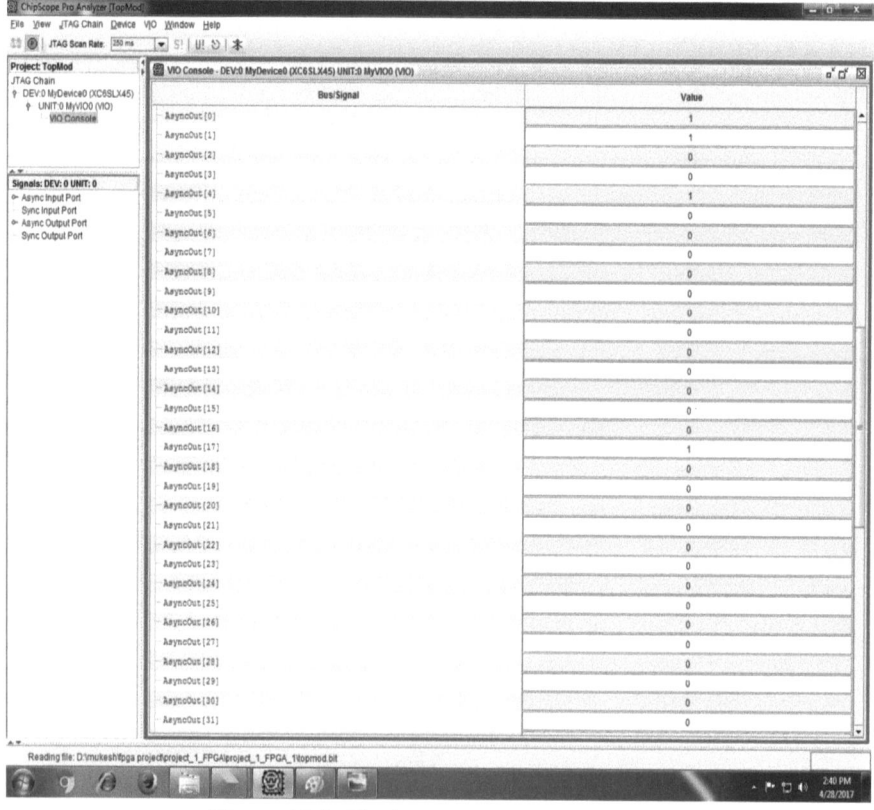

Figure 11.2: The input to control FPGA kit

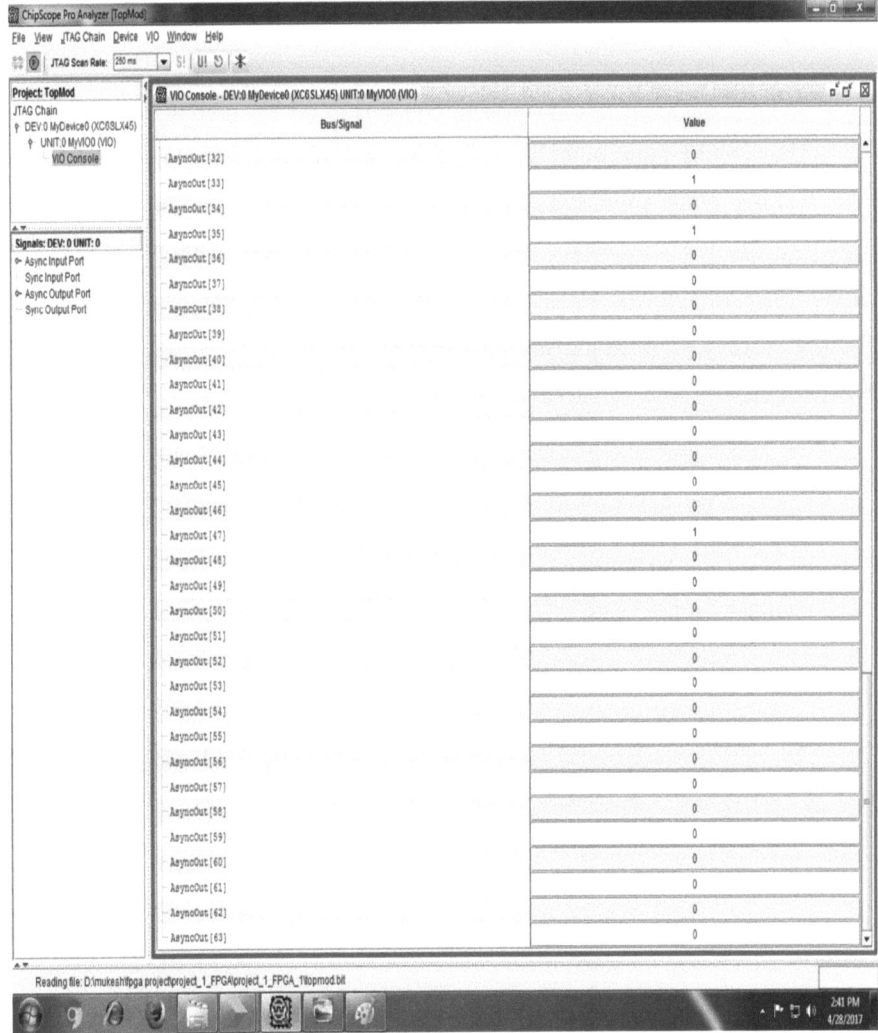

Figure 11.3: The output to control FPGA kit

Here the input and output of asynchronous enable pin gives the result with the control of FPGA. The output from top module result is given to the input of FPGA kit. The output of FPGA kit with control the result is given to the input of top module. Thus, VHDL simulation is verified and implemented in FPGA kit.

CHAPTER 12

CONCLUSION

The floating point multiplier is varied by 32 bit inputs. The simulation for both floating point shift and add multiplier and booth multiplier design are implemented and analyzed. The more accurate results are performed by eliminating the rounding modes and normalization in shift and add floating point multiplier. Comparing the booth multiplication, the delay and silicon area are 20% lower in the shift and add floating point multiplier. Also, the total power of shift and add floating point multiplier is 18% reduced when compared to the total power of booth multiplication. Thus, the shift and add floating point multiplier having better performs than the booth multiplication. The whole synthesis and simulation were done using VHDL/Verilog HDL and Cadence EDA tool and also the implementation in FPGA kit is performed.

REFERENCES

[1] Ms.Radhika Jumde, Ms.Gauri Jambhulkar, Ms.Megha Chalakh, Ms.Dhanashri Bhagat, "32 bit Single Precision floating point Multiplier", *International Engineering Journal For Research & Development,* Vol. 2, No. 3, pp. 1-5.

[2] Earl E. Swartzlander Jr., Hani H.M. Saleh, "A floating point fused dot product unit", *IEEE Conference*, New York, Volume- 61, No. 2, pp.427-431, February 2012.

[3] U.V.Chaudhari and Prof.A.P.Dhande, "Design and Simulation of Binary floating point multiplier using VHDL", *International Journal of Scientific & Engineering Research*, Vol. 6, No. 2, pp. 9-12, February-2015.

[4] Sharmila Hemanandha, Siva Subramanian, "IEEE 754 floating point fused add sub unit", *ARPN Journal of Engineering and Applied Sciences*, Vol. 10, No. 14, pp.5730-5734, August-2015.

[5] S.Kishore, S.P.Prakash, "Floating point fused dot –Product Unit", *International Journal of Innovative Research in Science, Engineering and Technology,* Vol. 4, No. 6, pp.575-580, May-2015.

[6] V.Narasimha, V.Swathi, "Normalization on floating point Multiplication using Verilog HDL", *International Journal of VLSI and Embedded Systems*, Vol. 4, pp.586-591, August 2013.

[7] Sumi M.S., Sobin Daniel, "Multiplication of floating point numbers using VHDL", *International Journal of Engineering and Innovative Technology,* Vol. 4, No. 3, pp.185-188, September 2014.

[8] P.Gayatri, P.Krishna Kumari, V.Vamsi Krishna, T.S.Trivedi, V.Nancharaiah, "Design of Floating Point Multiplier Using VHDL", *International Journal of Engineering Research and Development*, Vol. 10, No. 3, pp.73-78, March 2014.

[9] Shanmugapriya.K, Ruth Anita Shirley.D, Prabu Venkateswaran.S, "Design and implementation of booth multiplier in comparison with other multipliers", *International Journal of Advance Research in Science and Engineering*, Vol. 4, No. 2, pp.351-359, February 2015.

[10] K.Sreenath, K. Shashidhar, "The Design of Low Power and High Speed Configurable Booth Multiplier", *International journal of VLSI system,* Vol. 2, No. 10, pp.1085-1088, November-2014.

[11] Wai-Leong Pang, Kah-Yoong Chan, Sew-Kin Wong, Choon- Siang Tan, "VHDL Modeling of floating point for VLSI designer Library", Vol. 12, No. 12, pp.678-688, December 2013.

[12] Kishore Shinde, A.K. Kuresh-RISET, "Hardware implementation of configurable booth multiplier on FPGA", *49th IRF International Conference,* Pune, India, Vol. 46, No. 8, pp.60-63, 21st February 2016.

[13] Megha jain, Pallavee Jaiswal.y, "Implementation of Booths Algorithm i.e Multiplication of Two 16 Bit Signed Numbers using VHDL and Concept of Pipelining", *International Research Journal of Engineering and Technology*, Vol. 03, No. 06, pp.492-496, June-2016.

[14] Soniya, Suresh Kumar, "A Review of Different Type of Multipliers and Multiplier-Accumulator Unit", *International Journal of Emerging Trends & Technology in Computer Science*, Vol. 2, No. 4, pp.364-368, July – August 2013.

APPENDIX

[1] Code for Top module:

Library ieee;

use IEEE.STD_LOGIC_1164.all;

entity top_1 is

port(

x : in STD_LOGIC_VECTOR(31 downto 0);

y : in STD_LOGIC_VECTOR(31 downto 0); carry : out STD_LOGIC;

z : inout STD_LOGIC_VECTOR(31 downto 0));

end top_1;

architecture top_level_arc of top_1 is

component exponent is

port(

a,b : in STD_LOGIC_VECTOR(7 downto 0);

-- bias : in STD_LOGIC_VECTOR(7 downto 0); carry : out STD_LOGIC;

sum1 : out STD_LOGIC_VECTOR(7 downto 0));

end component;

component mantissa is

port(

a : in STD_LOGIC_VECTOR(22 downto 0);

b : in STD_LOGIC_VECTOR(22 downto 0);

p : out STD_LOGIC_VECTOR(47 downto 0));

end component;

```vhdl
component sign1 is
port (a : in STD_LOGIC;
b : in STD_LOGIC;
sign : out STD_LOGIC);
end component;
signal h : std_logic_vector (47 downto 0);
signal w,u : std_logic_vector (7 downto 0);
begin
u(7 downto 0)<= y(30 downto 23);
w(7 downto 0)<= x(30 downto 23);
t1 : sign1    port map (x(31),y(31),z(31));
t2 : exponent port map (u(7 downto 0),w(7 downto 0),carry,z(30 downto 23));
t3 : mantissa port map (x(22 downto 0),y(22 downto 0),h(47 downto 0));
z(22 downto 0)<= h(47 downto 25);
end top_level_arc;
Library ieee;
use IEEE.STD_LOGIC_1164.all;
entity adder_8ibit is
port(a : in STD_LOGIC_VECTOR(7 downto 0);
b : in STD_LOGIC_VECTOR(7 downto 0);
carry : out STD_LOGIC;
sum : out STD_LOGIC_VECTOR(7 downto 0));
end adder_8ibit;
```

```vhdl
architecture adder_8ibit_arc of adder_8ibit is

Component fa1 is

port (a : in STD_LOGIC;

b : in STD_LOGIC;

c : in STD_LOGIC;

sum : out STD_LOGIC;

carry : out STD_LOGIC);

end component;

signal s : std_logic_vector (6 downto 0);

begin

u0 : fa1 port map (a(0),b(0),'0',sum(0),s(0));

u1 : fa1 port map (a(1),b(1),s(0),sum(1),s(1));

u2 : fa1 port map (a(2),b(2),s(1),sum(2),s(2));

u3 : fa1 port map (a(3),b(3),s(2),sum(3),s(3));

u4 : fa1 port map (a(4),b(4),s(3),sum(4),s(4));

u5 : fa1 port map (a(5),b(5),s(4),sum(5),s(5));

u6 : fa1 port map (a(6),b(6),s(5),sum(6),s(6));

u7 : fa1 port map (a(7),b(7),s(6),sum(7),carry);

end adder_8ibit_arc;

Library ieee;

use IEEE.STD_LOGIC_1164.all;

use IEEE.STD_LOGIC_UNSIGNED.all;

entity mantissa is
```

```vhdl
port(a : in STD_LOGIC_VECTOR(22 downto 0);
b : in STD_LOGIC_VECTOR(22 downto 0);
p : out STD_LOGIC_VECTOR(47 downto 0));
end mantissa;
architecture multipier1_arc of mantissa is
begin
process(a,b)
variable pv,bp:STD_LOGIC_VECTOR(47 downto 0);
variable k,l:STD_LOGIC_VECTOR(23 downto 0);
begin
k := "1" & a;
l := "1" & b;
pv:="000000000000000000000000000000000000000000000000";
bp:= "000000000000000000000000"& l;
for i in 0 to 23 loop
if k(i) = '1' then
pv:=pv+bp;
end if;
bp:= bp(46 downto 0)&'0';
p<=pv;
end loop;
end process;
end multipier1_arc;
```

Library ieee;

use IEEE.STD_LOGIC_1164.all;

entity fa1 is

port (a : in STD_LOGIC;

b : in STD_LOGIC;

c : in STD_LOGIC;

sum : out STD_LOGIC;

carry : out STD_LOGIC);

end fa1;

architecture fa1_arc of fa1 is

begin

sum <= a xor b xor c;

carry <= (a and b) or (b and c) or (c and a);

end fa1_arc;

Library ieee;

use IEEE.STD_LOGIC_1164.all;

entity exponent is

port(a : in STD_LOGIC_VECTOR(7 downto 0);

b : in STD_LOGIC_VECTOR(7 downto 0);

-- bias : in STD_LOGIC_VECTOR(7 downto 0);carry : out STD_LOGIC;

sum1 : out STD_LOGIC_VECTOR(7 downto 0);

end exponent;

architecture adder_ibit_arc of exponent is

```vhdl
component adder_8ibit is
port(a : in STD_LOGIC_VECTOR(7 downto 0);
b : in STD_LOGIC_VECTOR(7 downto 0);    carry : out STD_LOGIC;
sum : out STD_LOGIC_VECTOR(7 downto 0));
end component;
signal s1 : std_logic_vector (8 downto 0);
begin
n0 : adder_8ibit port map (a(7 downto 0),b(7 downto 0),s1(8),s1(7 downto 0));
n1 : adder_8ibit port map ("10000001",s1(7 downto 0),carry,sum1(7 downto 0));
end adder_ibit_arc;
Library ieee;
use IEEE.STD_LOGIC_1164.all;
entity sign1 is
port (a : in STD_LOGIC;
b : in STD_LOGIC;
sign : out STD_LOGIC);
end sign1;
architecture sign_arc of sign1 is
begin
sign <= a xor b;
end sign_arc;
```

[2] Code for Booth Multiplier:

Library ieee;

use IEEE.STD_LOGIC_1164.all;

entity top2 is

port(x : in STD_LOGIC_VECTOR(31 downto 0);

y : in STD_LOGIC_VECTOR(31 downto 0); carry : out STD_LOGIC;

z : inout STD_LOGIC_VECTOR(31 downto 0));

end top2;

architecture top_level_arc of top2 is

component exponent is

port(a,b : in STD_LOGIC_VECTOR(7 downto 0);

--bias : in STD_LOGIC_VECTOR(7 downto 0); carry : out STD_LOGIC;

sum1 : out STD_LOGIC_VECTOR(7 downto 0));

end component;

component booth_multiplier is

port(x, y: in std_logic_vector(23 downto 0);

O: out std_logic_vector(47 downto 0));

end component;

component sign1 is

port (a : in STD_LOGIC;

b : in STD_LOGIC;

sign : out STD_LOGIC);

end component;

signal h : std_logic_vector (47 downto 0);

signal w,u : std_logic_vector (7 downto 0);

begin

u(7 downto 0)<= y(30 downto 23);

w(7 downto 0)<= x(30 downto 23);

t1 : sign1   port map (x(31),y(31),z(31));

t2 : exponent port map (u(7 downto 0),w(7 downto 0),carry,z(30 downto 23));

t3 : booth_multiplier port map ("1"& x(22 downto 0),"1"& y(22 downto 0),h(47 downto 0));

z(22 downto 0)<= h(47 downto 25);

end top_level_arc;

Library ieee;

use IEEE.STD_LOGIC_1164.all;

entity adder_8ibit is

port(a : in STD_LOGIC_VECTOR(7 downto 0);

b : in STD_LOGIC_VECTOR(7 downto 0);

carry : out STD_LOGIC;

sum : out STD_LOGIC_VECTOR(7 downto 0));

end adder_8ibit;

architecture adder_8ibit_arc of adder_8ibit is

Component fa1 is

port (a : in STD_LOGIC;

b : in STD_LOGIC;

c : in STD_LOGIC;

sum : out STD_LOGIC;

carry : out STD_LOGIC);

end component;

signal s : std_logic_vector (6 downto 0);

begin

u0 : fa1 port map (a(0),b(0),'0',sum(0),s(0));

u1 : fa1 port map (a(1),b(1),s(0),sum(1),s(1));

u2 : fa1 port map (a(2),b(2),s(1),sum(2),s(2));

u3 : fa1 port map (a(3),b(3),s(2),sum(3),s(3));

u4 : fa1 port map (a(4),b(4),s(3),sum(4),s(4));

u5 : fa1 port map (a(5),b(5),s(4),sum(5),s(5));

u6 : fa1 port map (a(6),b(6),s(5),sum(6),s(6));

u7 : fa1 port map (a(7),b(7),s(6),sum(7),carry);

end adder_8ibit_arc;

library ieee;

use ieee.std_logic_1164.all;

use ieee.numeric_std.all;

use ieee.std_logic_unsigned.all;

entity booth_multiplier is

port(x, y: in std_logic_vector(23 downto 0);

O: out std_logic_vector(47 downto 0));

end booth_multiplier;

```vhdl
architecture booth_multiplier_arc of booth_multiplier is
begin
process(x, y)
variable a: std_logic_vector(48 downto 0);
variable s,p : std_logic_vector(23 downto 0);
variable i:integer;
begin
a := "0000000000000000000000000000000000000000000000000";
s := y;
a(24 downto 1) := x;
for i in 0 to 23 loop
 if(a(1) = '1' and a(0) = '0') then
  p := (a(48 downto 25));
  a(48 downto 25) := (p - s);
 elsif(a(1) = '0' and a(0) = '1') then
  p := (a(48 downto 25));
  a(48 downto 25) := (p + s);
 end if;
 a(47 downto 0) := a(48 downto 1);
end loop;
O(47 downto 0) <= a(48 downto 1);
if (x(23)='1' or y(23)='1')then
 O(47 downto 0)<= std_logic_vector(unsigned(not a(48 downto 1) + '1'));
```

end if;

end process;

end booth_multiplier_arc;

Library ieee;

use IEEE.STD_LOGIC_1164.all;

entity fa1 is

port (a : in STD_LOGIC;

b : in STD_LOGIC;

c : in STD_LOGIC;

sum : out STD_LOGIC; carry : out STD_LOGIC);

end fa1;

architecture fa1_arc of fa1 is

begin

sum <= a xor b xor c;

carry <= (a and b) or (b and c) or (c and a);

end fa1_arc;

Library ieee;

use IEEE.STD_LOGIC_1164.all;

entity exponent is

port(a : in STD_LOGIC_VECTOR(7 downto 0);

b : in STD_LOGIC_VECTOR(7 downto 0);

-- bias : in STD_LOGIC_VECTOR(7 downto 0); carry : out STD_LOGIC;

sum1 : out STD_LOGIC_VECTOR(7 downto 0));

end exponent;

architecture adder_ibit_arc of exponent is

component adder_8ibit is

port(a : in STD_LOGIC_VECTOR(7 downto 0);

b : in STD_LOGIC_VECTOR(7 downto 0);   carry : out STD_LOGIC;

sum : out STD_LOGIC_VECTOR(7 downto 0));

end component;

signal s1 : std_logic_vector (8 downto 0);

begin

n0 : adder_8ibit port map (a(7 downto 0),b(7 downto 0),s1(8),s1(7 downto 0));

n1 : adder_8ibit port map ("10000001",s1(7 downto 0),carry,sum1(7 downto 0));

end adder_ibit_arc;

Library ieee;

use IEEE.STD_LOGIC_1164.all;

entity sign1 is

port (a : in STD_LOGIC;

b : in STD_LOGIC;

sign : out STD_LOGIC);

end sign1;

architecture sign_arc of sign1 is

begin

sign <= a xor b;

end sign_arc;